THE POWER OF THE PRESS

THE STORY OF EARLY CANADIAN PRINTERS AND PUBLISHERS

THE POWER OF THE PRESS

THE STORY OF EARLY CANADIAN PRINTERS AND PUBLISHERS

CHRIS RAIBLE

JAMES LORIMER & COMPANY LTD., PUBLISHERS
TORONTO

James Lorimer & Company Ltd. acknowledges the support of the Ontario Arts Council. We acknowledge the support of the Government of Canada through the Book Publishing Industry Development Program (BPIDP) for our publishing activities. We acknowledge the support of the Canada Council for the Arts for our publishing program. We acknowledge the support of the Government of Ontario through the Ontario Media Development Corporation's Ontario Book Initiative.

Library and Archives Canada Cataloguing in Publication

Raible, Chris, 1933-
The power of the press : the story of early Canadian printers and publishers / Chris Raible

Includes bibliographical references and index.
ISBN 978-1-55028-982-4

1. Press — Canada — History. 2. Printing--Canada — History. 3. Press and politics — Canada — History. I. Title.

PN4907.R35 2007 071'.109034 C2007-900337-0

Illustration on Contents page: pieces of type

James Lorimer & Company Ltd., Publishers
317 Adelaide Street West, Suite 1002
Toronto, ON
M5V 1P9
www.lorimer.ca

Printed and bound in Canada

Contents

Acknowledgements

To Don Ritchie, long-time printer at Mackenzie House, Toronto, for first prompting my interest in the history of Canadian letterpress printing.

To Lou Cahill and all at the Mackenzie Printery and Newspaper Museum in Queenston, Ontario, for encouraging my continuing research of William Lyon Mackenzie as a printer-editor.

To Steven Sword of Stouffville, Ontario, and Joe Landry at the Nova Scotia College of Art and Design for sharing their extensive knowledge of letterpress printing technology.

To Peter Cazaly, Nancy Morrell and Ron Lefebvre at Upper Canada Village, Derek Cooke at Black Creek Pioneer Village, and Fiona Lucas and Alex Sawa at Mackenzie House, Toronto, for making possible photographs at their sites; and to Jackie MacRae and Rob Skeoch, photographers.

To all the curators and administrators of museums and historic sites across Canada for supplying information on the letterpress equipment in their collections.

To Lynn Schellenberg, Catherine MacIntosh, and Pamela Martin, editors at James Lorimer & Company, for their professional expertise.

And especially to Pat, without whose love, encouragement, knowledge, advice, patience, sensitivity and judgement, nothing worth reading would ever be written by me.

The Press is the life, the safeguard, the very heart's blood of a free country; the test of its worth, its happiness, its civilization.

— William Lyon Mackenzie, *Colonial Advocate*, January 26, 1832

Introduction

This survey of early Canadian printing is not meant to be comprehensive. It is a brief sketch of technological changes and their relation to social and political developments. This survey also sketches the careers of some of the printers who were important to the development of Canadian printing. Their stories are illustrative of the work and worries of the many other early printer-editors who laid the groundwork for the industry in

Left: A compositor lifts a locked forme of types that is now ready for the press.
Above: The Mackenzie Printery and Newspaper Museum is the former home and office of William Lyon Mackenzie, an important figure in the history of printing in Canada.

A proof is read for typographic errors and possible editorial changes before final printing.

Canada — some well-known, others less so, mostly men, but also a few women.

When printers first began picking types and pulling presses in Canada in the mid-eighteenth century, they were using a process, typography, that had changed very little since it was first invented by Johannes Gutenberg in Germany three hundred years earlier.

Before the development of printing, all documents were manuscripts — written, copied and re-copied by hand. Because of the time and expense involved, books were scarce and newspapers were non-existent.

Around 1440, Gutenberg and his colleagues invented a technology to duplicate texts mechanically. (Korean printers had been printing with movable type for more than a century before Gutenberg, but their technology never crossed the divide between east and west.) For some twenty years Gutenberg and his associates experimented, labouring to design types, devise molds, and discover the best lead alloys. They invented methods for arranging pieces of type and holding them tightly together for pressing. They adapted wine and other presses to make printing presses. They tried different inks and papers. Once perfected, their printing process made possible a highly successful commercial enterprise. It required skilled workers to set types, to press paper, and to cut and collate sheets to make the finished product. But the process itself was a way to make multiple copies more accurately, comparatively quickly, and at a much lower cost than the pen-and-ink work of scribes.

Gutenberg's revolutionary invention put words on paper in an entirely new way. It used types, small metal blocks, each one topped with the mirror image of a letter. These individual letters could be assembled into words, sentences, paragraphs, and pages. Printing from wooden blocks carved with words and images was by then a common art, but Gutenberg's blocks were different — thousands of distinct pieces. Every letter of the alphabet, and every punctuation mark or symbol, was cast in lead hundreds of times. Each letter was identical to its original pattern — every letter *a* looked like every other *a*. These separate pieces of type were arranged, letter by letter, into words, sentences, and paragraphs. The entire surface of the type was then inked and pressed to paper to produce a printed page. Re-inking and pressing made more copies, all from the same set types. And the types were re-usable — when the task was complete, the types were cleaned and sorted, ready for the next job.

Gutenberg is best known for a printed edition of Scripture, often called his "42-line Bible" (each page had forty-two lines of print

Rolling a brayer on an inkstone covers its surface with ink.

Wooden cases of types: capital letters were stored in the upper case, small letters in the lower case.

in two columns). But his print shop also produced calendars, indulgences, and pamphlets — ephemeral items, forerunners of what came to be called "job printing," inexpensive printed materials that were not intended to be permanent. Furthermore, he could print on demand, producing copies comparatively quickly. Soon printers were using Gutenberg's new technology to publish journals and newspapers.

Thanks to Gutenberg and other German printers, printing spread quickly. By 1476 William Caxton was printing in England. Printers were starting up businesses in city after city — by 1500 there were more than 250 print shops in England and Western Europe.

The earliest printers, all the way back to Gutenberg, mainly published books: Bibles, scientific treatises, law books, and other volumes. In Canada, however, the first printers spent most of their time putting out newspapers as vehicles for the widespread, immediate communication of reports, notices, advertisements, opinions, and news. Unlike the private correspondence of individuals, these papers

An assistant inks a forme before the pressman lowers the tympan to print a sheet of paper.

were public documents. Paradoxically, it is their currency and ephemeral nature that makes these old papers so valuable historically.

The invention of printing affected religion, philosophy, science, medicine, education, government, commerce, travel, literature, music — nearly every aspect of human endeavour enlarged and flourished. Printed pamphlets and books were easy to produce, distribute, buy, and read. Printed works moved freely from town to town, country to country, continent to continent. Ideas and information became more difficult to restrict, repress, or censor — an author might be silenced, imprisoned, even burned for a work deemed treasonous, blasphemous, or heretical, but printed copies survived and continued to circulate.

The author of *Typographia*, an influential early nineteenth-century manual for printers,

A printed sheet is carefully peeled from the press.

waxed poetical in contrasting his time with the dark ages before printing:

> *By means of the press, curiosity is roused; the mind is expanded; it no longer groans under the pressure of Ignorance and Folly — Vice and Virtue are depicted in their true colours; and Cruelty and Oppression are ever held up to the scorn and detestation of the world: in a word the harvest is now complete.*

Within a hundred years of Gutenberg, Juan Pablos was pulling a press in Mexico. In another hundred years, Samuel Daye had a print shop in Massachusetts. By the early 1700s, printers were at work in Philadelphia, New York, and Boston, where apprentice printer Benjamin Franklin was starting out on his long public career.

At mid-century, 1751, Bartholomew Green arrived in Halifax to start up Canada's first printing business.

1 The Politics of Printing

The King's Printers, 1751–1800

Printing was first brought to Canada by the British colonial government. (In its more than two-hundred-year history, there were no printing presses in New France.) Two years after establishing a garrison and colonizing Nova Scotia, the government lured a printer, Bartholomew Green, from Boston to Halifax. Green died within weeks of his arrival, but he was quickly replaced by his former partner John Bushell, who made the move to Halifax to become the colony's first "King's Printer."

The post was not a full-time position; Bushell was no privileged placeman. He was given a warm, dry place to work and some money for expenses, but no salary. He was an independent entrepreneur with his own press and types. He paid his own bills, purchased his own supplies, and collected money for each printing assignment he undertook. In time, he hired employees, kept accounts, and engaged in all the activities of a small business. But his main source of income was government printing.

A printing press was an essential tool of colonial administration — "A printer is indispensably necessary," Lieutenant-Governor John Graves Simcoe would later insist, as, in 1791,

A selection of wooden type used to give variety to a printed page.

King's Printers owned their own types and machinery (and would never store their types in this jumbled fashion).

he made plans for the new province of Upper Canada. In the early decades of British North America, the primary purpose of the press was to propagate official propaganda — freedom of the press was an alien idea. Colonial administrators needed a medium for their messages. Formal proclamations, laws, and regulations had to be made public as quickly and as clearly as possible. Prior to the availability of printers, clerks wrote copies by hand, at a price per word. Printed copies were not only cheaper, they could be corrected before the final printing. They were also easier to read and, more importantly, they were more readily given wide distribution.

Printing ensured executive efficiency. Officials were expected to circulate copies of all official records, all court rulings, and all the statutes enacted annually by colonial legislatures. Pre-printed forms, standard texts with blanks for filling in specific details, also made life easier for administrators as they granted land, appointed men to office, purchased goods, hired employees, kept records, and engaged in myriad other matters.

Large types made of wood were used for posters and headlines.

But most of all, printing made possible an official newspaper. In every province, the columns of a weekly *Gazette* (named after the *London Gazette,* the English governmental organ since 1665) carried the many notices colonial administrators wanted to circulate. All government pronouncements, announcements, appointments, rules, and regulations were "gazetted," printed in the newspaper. By being published, that is, made public, they became official.

On March 23, 1752, Bushell published the first edition of the *Halifax Gazette*. With an apology for long delays, the printer declared:

> *[T]he Letter Press is now commodiously fixed for the Printing Business, all such Gentlemen, Merchants, and others, as may have Occasion for any Thing in that Way, may depend upon being served in a reasonable and expeditious Manner, by their Most Obedient, Humble Servant, John Bushell.*

However, this first Canadian newspaper was anything but an independent organ. Bushell printed all the official information his superiors wished to disseminate and filled the rest of his two-page paper with shipping news, items culled from American and British papers, and whatever paid advertising he could get.

Set lines of types in a galley are ready in the proofing press.

Following long-established printing practices, to print the *Gazette* Bushell used a variety of type in different sizes or fonts, every font consisting of tens of thousands of pieces, the quantity of each letter or character depended on its frequency of use — more for *e*, fewer for *k*. Each font was stored in its own case, a tray with many compartments to keep the letters separate.

To typeset the *Gazette*, or any printed matter, the printer held a composing stick in one hand and with the other selected types from propped-up cases. Letters picked one at a time formed words, line by line, until the stick was full. Set lines of types were then moved to a galley tray and the process would begin again until all the desired text was typeset. The galley of set types was then inked and a page printed at a small proofing press. The sample page was carefully proofread, then errors were corrected and editorial changes made by resetting the type.

Working on an imposing stone, a hard-surfaced table, he then laid out corrected blocks

Above: Quoins, both older-style wedges of wood and later, metal shapes, lock the laid-out type in a forme. Right: A tied block of types ready for later use.

of types in a chase, a metal frame somewhat larger than the final page to be printed. Margins and blank areas between sections of type were created with furniture, hardwood pieces in various sizes. Space limitations might require "cutting to the chase," that is, removing part of the text. Once a chase was laid out as desired, all the types and pieces of furniture were locked in place and the entire forme was ready for printing. Throughout the composing process, care had to be taken that a galley or forme was not dropped, reducing all the set types to pye, a confused jumble to be sorted and reset.

The completed forme was placed on the flat surface or bed of a press — "put to bed."

Picking types from a case to create lines of text in a composing stick.

Hinged to the bed were two large plates: the tympan, covered with parchment, and the frisket, an open frame. A sheet of moistened paper was laid on the tympan and the frisket was folded down to protect the paper's margins. Meanwhile, a second worker (usually an apprentice) applied ink to the surface of the type. The tympan was then lowered onto the forme and the bed was cranked into the body of the press itself. The pressman pulled a large lever to lower a heavy flat surface, the platen, pressing tympan to types and transferring ink to paper. The platen was then raised, the bed cranked out, the tympan lifted, and the print-

Pounding quoins in place to lock set types and furniture in a forme.

Standard wood or metal cuts, as well as borders, rule lines, and ornaments were obtained from the type foundry.

ed sheet removed and hung to dry, the process repeated to make the next copy.

The press that Bushell pulled to print his papers, a so-called "common" press, was probably manufactured in England. It was very like the press used by Gutenberg in the fifteenth century, or the eighteenth-century press on which Franklin learned his trade in Boston. Operating a common press was slow work. Because the platen was only half the size of the press bed, most jobs required double pulling. The bed would be cranked in halfway, the lever pulled to lower the platen and raise it again, the bed moved further into the press, and the lever pulled again. One man could print barely fifty sheets an hour — two men working together at top speed, up to two hundred an hour.

When the printing of a forme was complete, a worker (usually an apprentice) cleaned off the ink, unlocked the chase, and removed all the furniture. The types were then distributed,

Placing the paper on the tympan.

separated, and sorted back into the cases, each font into its proper case, each letter into its proper section. The distributor had to "mind his p's and q's" to be sure the type was correctly sorted, ready for a compositor to use again to set new words, lines, and pages.

Bushell's types were cast in England, almost certainly by the firm founded by William Caslon in 1722–23. Caslon types, or copies of them, were used by most printers on both sides of the Atlantic. (Many modern typefaces, including the fonts used in the printing of this book, continue to be influenced by Caslon's designs.) The *Gazette* masthead featured pictures of a sailing ship and a foot traveller — printed from cuts, carved blocks of wood. These two cuts were generic images that, like types, could be used and re-used. Later printers sometimes commissioned local craftsmen to carve cuts for particular print jobs.

The first Canadian King's Printer was soon in economic and political difficulties. Provincial administrators saw the *Gazette* as an instrument of the government. Its readers were their subjects. Its printed words were intended to inform and educate — and to direct and

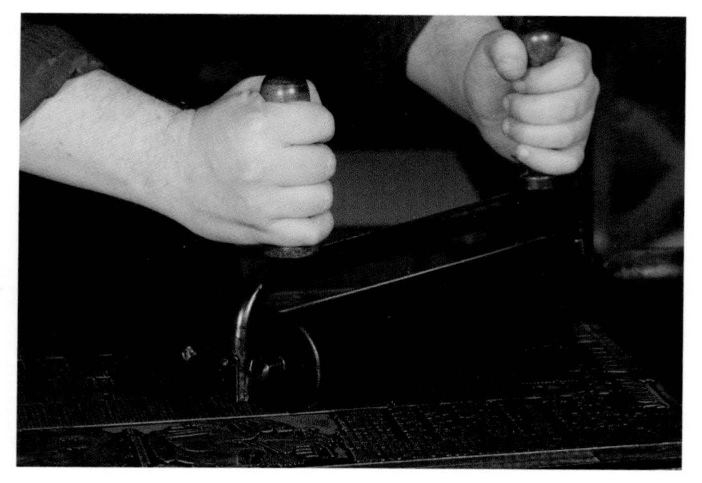

Above: Lowering the frisket.
Left: Applying ink to the types.

regulate. The purpose of the paper was not to persuade but to command.

The *Gazette* was subsidized by the government, but it was not completely underwritten. Bushell had to solicit paid advertising and sell subscriptions for it to be a paying proposition. Otherwise he could not afford to print it at all. Unfortunately, what government administrators thought their subjects needed and what Bushell's subscribers and advertisers wanted

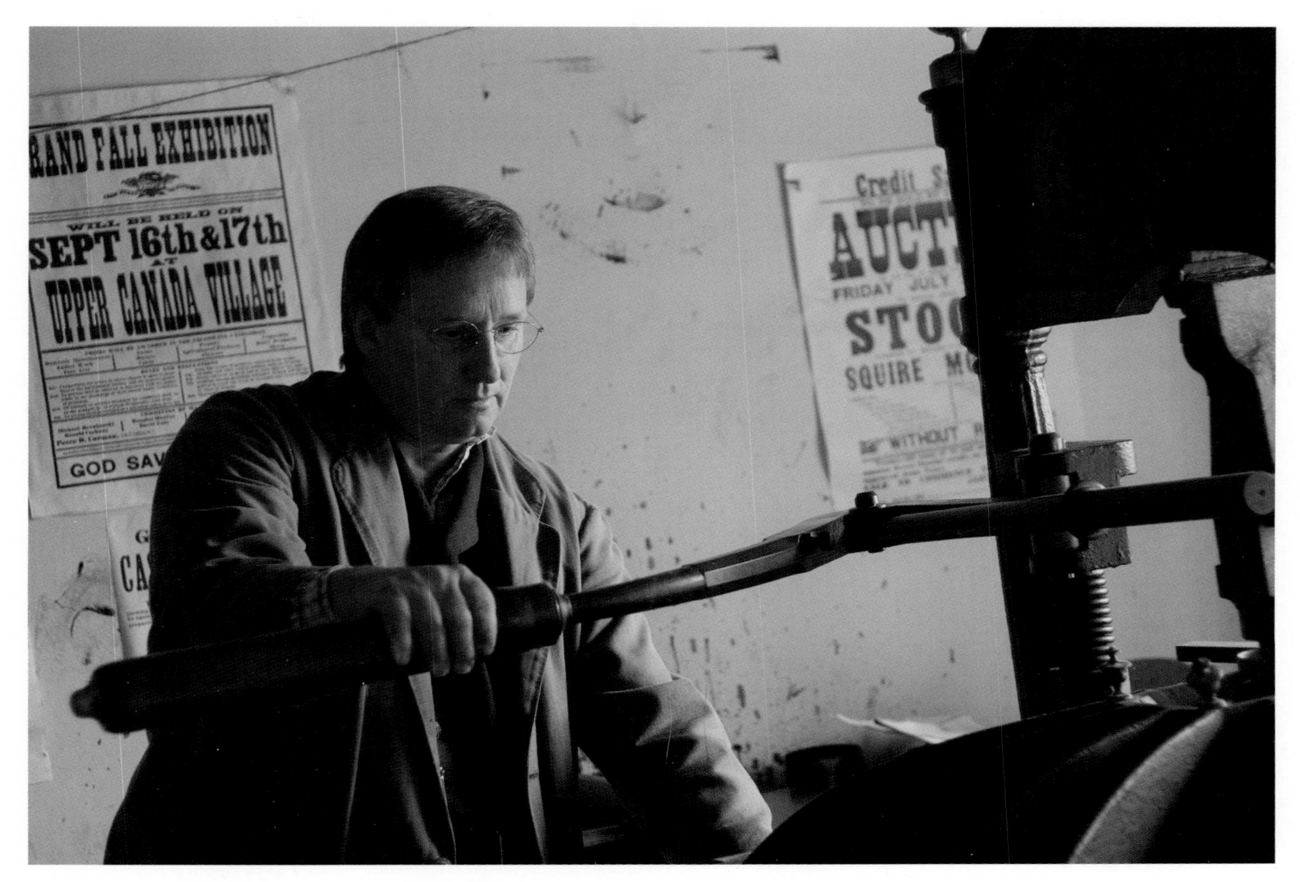

Pulling a large lever lowers the platen, pressing paper to inked types.

were not necessarily the same. The King's Printer was thus caught between the demands of his superiors and the interests of his readers.

And Bushell had another problem. Even though his editorial freedom was limited, his government employers continued to have lingering doubts about his loyalty. In those years just prior to the American Revolution, an American-born printer was suspect. The Provincial Secretary took over editing the *Gazette*, leaving the printing to Bushell.

He also had personal problems. Printing may have been indoor work with no heavy lifting, an occupation largely unaffected by changes of season and variations of weather, but it was often stressful. His daughter, Elizabeth, assisted with composing and even press work, but long hours of concentrated yet tedious activity took their toll on him. He muddled his accounts and was prosecuted for debt. And his ordeals were aggravated by an addiction to alcohol, which led to his death in

Peeling a printed sheet from the forme.

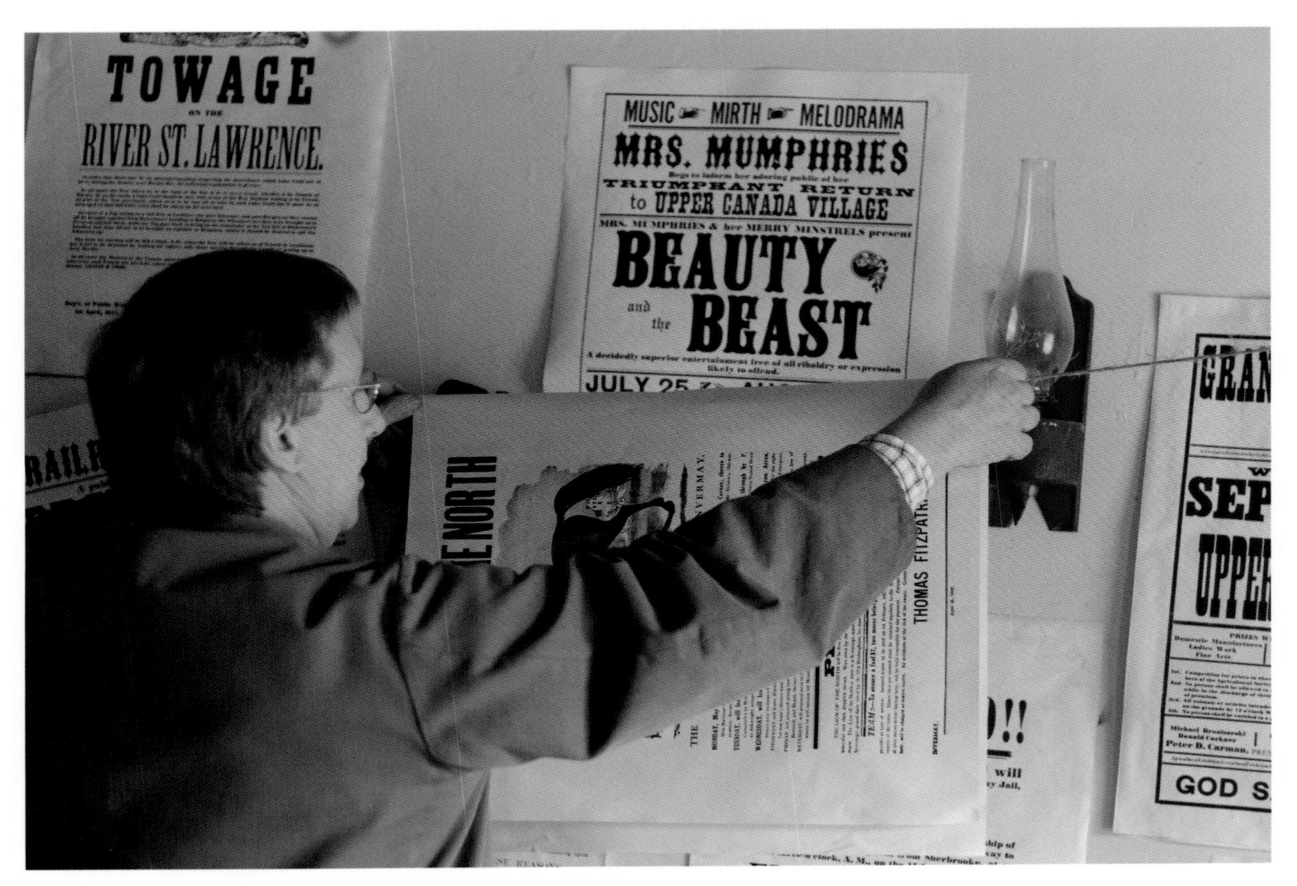

Freshly printed sheets are hung to dry.

1761: Canada's first incident of a problem that constantly beset the whole trade — drink was the curse of the printing class.

Over the course of the next few decades, the printed word had an impact on nearly every aspect of community life. Ease of communication opened new opportunities for spreading ideas and for making money. Colonial newspapers became instruments of colonial commerce. In Halifax — and soon thereafter in other colonies — the columns of the *Gazette* were employed by merchants promoting wares, innkeepers soliciting custom, lawyers posting notices, organizations publicizing meetings, sellers setting auctions, employers offering jobs, craftsmen seeking work, tradesmen soliciting business, and farmers searching for lost cattle or stolen horses. Advertisers also addressed the public with broadsides, single sheets printed as handbills or posters.

Some private printing was less ephemeral. In time, printers in Canada began producing

Printing broadsides — posters and handbills — provided much of a printer's income.

books and pamphlets. Some were religious: sermons, catechisms, and hymnals. Some were inspirational or educational: poems, political tracts, and didactic treatises. Some were commercial: directories, catalogues, and manuals. More mundane job work, printing a wide range of items, from cards and labels to receipts and indentures, occupied much of a printer's time and produced much of his income.

The second King's Printer in Nova Scotia, Anton Heinrich, learned his trade in his native Germany, but came to America as a fifer in the British army. Following the siege of Louisbourg in 1758, he left the army and returned to printing, first in New Jersey and then in Halifax. He acquired Bushell's enterprise, including publishing the *Gazette*, after anglicizing his name to the more politically acceptable Anthony Henry. (Heinrich also published a German-language paper for the German settlers in nearby Lunenburg under his original name.) For several years he prospered, but in October 1765 he found himself in deep trouble for printing an editorial in the *Gazette* suggesting that Nova Scotians were opposed to the Stamp Act. This was a law taxing all paper used for newspapers, legal documents, licences, and the like — every sheet was required to bear an official stamp certifying that the duty had been paid. Enacted by the British Parliament, the Stamp Act became a notorious cause of discontent in the American colonies. Henry blamed the offending paragraph on his apprentice, Isaiah Thomas, but doubts about Henry's loyalty persisted. Much worse, an edition of the *Gazette* was printed on paper stock lacking the required stamps. Officials were outraged. Thomas fled back home to Massachusetts, where he went on to become a prominent publisher, and later an historian of printing. Henry protested his innocence but the government withdrew its patronage and the *Gazette* closed down. Henry launched his own independent newspaper, but Halifax was not big enough for two print shops. After a few years he made his peace with the administration, trading independence for income. Eventually he was re-commissioned as King's Printer and his paper metamorphosed into the *Royal Gazette and Nova Scotia Advertiser.* The enterprise flourished — Henry had learned the value of avoiding controversy.

Meanwhile in the city of Quebec in 1764, two Philadelphia printers, William Brown and Thomas Gilmore, launched the government-sponsored *Quebec Gazette/La Gazette de Québec*. The paper was bilingual, half English, half French. Although they printed official announcements as required, the editors had loftier aspirations. They wanted their *Gazette* to be "the most effectual Means of bringing about a thorough Knowledge of the *English* and *French* language to those of the two Nations now happily united in one in this Part of the World," publishing foreign news with "impartiality."

At a time when grievances were being voiced so vigorously by American colonists to the south, any suggestion of solidarity with their cause made colonial authorities in Canada uneasy. The Quebec printer-editors daringly added that their paper would "have nothing so much at heart, as the support of VIRTUE and MORALITY, and the noble cause of LIBERTY." Such ideas were dangerous.

For a time the *Quebec Gazette* enjoyed some degree of freedom, but the excessive cost of the infamous stamp tax forced the paper's suspension for several months. When the Stamp Act was rescinded in 1766, the editors resumed publication, resolving to be free from inspection or restrictions. In this period of intense controversy, they dared to reprint political commentary from Britain *and* from the American colonies. Official scrutiny increased and by 1770 the paper was being strictly censored by the government. Furthermore, the Brown and Gilmore enterprise experienced only limited financial success. In 1772, Gilmore, thirty-two years old, died of alcoholism, leaving Brown to carry on without him. Three years later, an American invasion of Canada stopped the presses altogether. Not until August 8, 1776, well after the Americans had been driven off, did Brown dare to go back into business, proclaiming to his readers that his paper, "justly merited the Title of THE MOST INNOCENT GAZETTE IN THE BRITISH DOMINIONS." He prospered until his death in 1789 when the print shop was taken over by his nephew Samuel Neilson. A few years later, his younger brother, John Neilson, became publisher of the *Quebec Gazette,* beginning a fifty-year career as an influential political journalist.

The early Halifax and Quebec printers had their political troubles, but these were small compared to those experienced by Fleury Mesplet in Montreal. Born and trained as a printer in France, Mesplet emigrated to Philadelphia where he was befriended by Benjamin Franklin. In the wake of the 1775 American occupation of Montreal, the French printer was enlisted to help rouse the support of the French Canadians to the cause of North American independence. The next spring, Mesplet arrived in Montreal bearing a Continental Congress commission and a borrowed press and types. A month later the Americans abruptly withdrew from Canada. Back in charge, the British authorities threw Mesplet in jail, even though he had had no time to actually print anything.

He was soon released; he had done nothing wrong. He ran a job printing shop for a couple of years, until 1778, when he took on the risk of publishing a newspaper, *La Gazette du commerce et littéraire, pour la ville et district de Montréal*, Canada's first wholly French newspaper. Mesplet was the printer; the editor was Valentin Jautard, an admirer of the radical ideas of Voltaire and French republicanism. In response to the radical tone set by Jautard, Colonial officials killed the *Gazette* by imprisoning both men. This time Mesplet languished in jail for three years. But he was the only competent printer in town, so in 1782, although technically still under arrest, he was released and permitted to go back into business.

In 1785, Mesplet hired a new editor and launched a bilingual newspaper, the *Montreal Gazette/La Gazette de Montréal.* Despite its name, this (Mesplet's second) *Gazette* was anything but an official organ, often critical of the clergy and the feudalism of Quebec society. When the French revolution erupted, the paper expressed fervent approval. But when France declared war against Great Britain in 1793, the *Gazette* quickly switched its allegiance and continued publication. Mesplet died eight months later, nearly bankrupt, but the *Gazette* carried on and is still published today.

Blocks of wood, furniture, were used to fill blank spaces in a forme. Thin wood strips, reglets, separated lines of types.

The first King's Printer in Upper Canada, Louis Roy, learned his trade in Quebec under William Brown's tutelage and later found work in Montreal with Mesplet. Recruited to come to Upper Canada, he set up shop in Newark (Niagara-on-the-Lake) printing the Lieutenant-Governor's speeches and proclamations. On April 18, 1793, Roy launched *The Upper Canada Gazette or American Oracle,* as it was resoundingly titled, promising to make "every exertion … so as to render the paper useful, entertaining, and instructive," that it might become, "the Vehicle of Intelligence in this growing Province, of whatever may intend to its internal benefit and common advantage."

Like King's Printers in other provinces, Roy soon found himself pulled in opposite direc-

A broadside posted in a tavern lists prices of the food and drink sold.

tions. The editor's employers were government officials, responsible to the British Crown. The *Gazette*, one of the tools with which they ruled, was their means for communicating their proclamations, regulations, and notices. These administrators were not in Canada for life. They expected to make their fortunes and then go home. Their interests were essentially British — news to them was British news. Most of Roy's readers, however, were Americans, permanently settled in the province, but still American in outlook with relatives and friends back in the United States. For them, American events and politics were of more interest than British, and, undoubtedly, exercised more influence.

These were times that tried officials' souls. The American Revolution was not long over and the threat of invasion was still very real. And England was at war with France. Nervous colonial administrators thought they sensed seditious tendencies in Roy's efforts, although they cited few specifics. Any issue of his paper

Printed on strong paper made from rags, newspapers could be passed from person to person, read and re-read.

that seemed to raise questions about government activities, cast doubt on the equity of government rulings, or neglected to publicize information important for the administration, was suspect. When Roy quoted — or perhaps misquoted — an address to the Grand Jury by Chief Justice Osgoode, the jurist ordered the printer never again "to mention my name without apprising me." The Lieutenant-Governor's wife complained that Roy was "a

Frenchman" who "cannot write good English."

By the autumn of 1793 Roy was seriously ill and under heavy criticism, this time not for what he had printed, but what he had failed to print. A year after the new province's first parliamentary session, Chief Justice Osgoode complained to Lieutenant-Governor Simcoe,

> *We are put to great Inconvenience for want of publication of the Acts of Last Session … I have interfered more than becomes me … Every day [Roy] furnishes a fresh Excuse … I have presumed to threaten dismission tho I believe without chance of better effect than any other Remonstrance.*

Newark, the provincial capital, was hardly a printer's paradise. Most of the inhabitants were illiterate British regular soldiers. Government officials were mostly occupied in competing for influence and applying for grants of land that might make them rich. Roy felt isolated and overworked. Whether he was fired or simply quit is not clear, but a year later he was gone.

Roy limped back to Montreal and established his own newspaper, the Montreal *Gazette*, in competition with the existing Montreal *Gazette* — thus, for a brief time, the city had two *Gazettes*, neither one official. But the climate, political and economic, was not hospitable and Roy moved on to New York where he died in 1799. Canadian authorities had long suspected him of being a closet republican. An obituary in the New York *Argus*, the last paper he worked for, offered confirmation.

> *His political principles were those of a Republican; and he laboured until his last fatal illness with all the zeal which the most ardent patriotism could inspire to promote the interest of Freedom's sacred cause.*

Roy's successors in Upper Canada fared little better. The next thirty years in the province would see seven more King's Printers — nearly all would face serious political troubles.

2 Freeing the Press

The Independent Printer-Editors, 1800–1850

The nineteenth century brought revolutionary changes to the Canadian printing business. King's Printers continued to produce for the government, but opportunities for serving a wider market were opening up. These, plus the availability of cheaper and more efficient presses, prompted a number of printers to start their own enterprises.

The first of these printers in Upper Canada, the brothers Gideon and Silvester Tiffany,

With improved technology, newspapers could be circulated more widely.

A single person often acted as both editor and printer.

started their own print shop in Newark after three stressful years as King's Printers. When first recruited by colonial administrators to replace Louis Roy, the New Yorkers were strictly instructed by the Lieutenant-Governor's secretary to use their "own good sense and discretion," and to print news "most favorable to the British Government." He who paid the printers was determined to set the tone.

The Tiffanys paid the admonition little heed. American news was what their readers wanted, so American news was what they printed. Chief Justice John Elmsley, labelling them "unprincipled and unattached republicans," complained that "every trifle relating to the damn'd States is printed in large character." Also, the printers were reprimanded for importing their paper from Albany, New York, instead of purchasing it from a mill in Montreal. Cross-border shopping, whether for goods or ideas, was forbidden.

When warnings failed to discipline the brothers, officials employed harsher measures. In April 1797, Gideon was formally charged with blasphemy. Exactly what prompted this prosecution is unclear, the trial record is incomplete. He was fired, fined £20, and jailed for a month. Soon thereafter Silvester was threatened with prosecution for "treasonable or seditious conduct." "It is a charge ungenerous and unfounded," he insisted. "God can bear the clearest testimony of my innocency." Again, the explicit grounds for the charge are not clear. Defending himself, Silvester declared that his allegiance was not to the province's chief administrator, but to a higher order. "As the people's printer, it is my duty to devote my head, heart and hands to their service ... The interests of the King and people are inseparable."

A new King's Printer, Titus Geer Simons, was appointed — one who knew nothing about printing. Simons was officially in charge, but the Tiffanys continued setting types and working the press (and determining much of the paper's content) until all provincial offices, including the *Gazette,* moved to York, the new provincial capital. With the government work gone Gideon went west and

Types were housed in wooden cases.

took up farming, but Silvester stayed on, and in 1799 launched the *Canada Constitution*, the province's first independent newspaper. A lack of subscribers and advertisers killed it in a few months. Silvester tried again a year later, resurrecting and renaming his paper the *Niagara Herald*, but fared no better. He moved back to New York State where there were more customers — and perhaps fewer critics.

Over the course of the next few decades, however, other independent printer-editors started enterprises in Upper Canada that were

Above: A completed forme ready for printing.
Left: Some printers continued to do short-run print jobs.

more successful. Two newspapers, the *Kingston Gazette* begun in 1810 and the Niagara *Gleaner* in 1817, prospered for many years. In 1819, the successful launch of a second paper, *Upper Canada Herald,* in Kingston, and a year later a second in York, the *Observer,* demonstrated an increasing demand for job printing and a growing appetite for news.

The printer-editors running these new, independent print shops were highly skilled craftsmen who set types accurately and attractively and operated presses quickly and efficiently. Free from government interference, they had only to please their customers and manage their employees — a journeyman or two, perhaps a clerk, and one or more apprentices.

It took years to become a proficient printer. A teenaged boy apprentice (no girls), often called a "printer's devil," began by cleaning types, learning to recognize the letter and font of each piece, and distributing, or sorting, the types back into the proper cases. Once he had memorized the layout of the separate compartments of the cases, sorting went quickly. The most used fonts were often in short supply and a compositor who ran out of a particular letter was "out of sorts." Once experienced at distributing, an apprentice proceeded to learn the skills of composing — setting, spacing, leading, proofing, and correcting. He also learned to mix and apply ink, to moisten and lay down paper, to work the press and to perform all the other operations involved.

An apprentice was formally indentured by his parents who also paid a fee for his training. He was provided with room and board, a modest clothing allowance, and a little spending money. The hours were long and the work taxing, but it required brain more than brawn, speed rather than strength. Legally bound for a period of four or five years, a runaway could be captured and returned to his master. In most printing offices, however, the camaraderie of fellow workers was high and fugitives were rare. At the completion of the course, an apprentice became a fully qualified journeyman.

It took several years for an apprentice to learn the trade and become a journeyman printer.

Independent printers, like King's Printers, earned much of their income from job printing. They also published a variety of books and pamphlets, some religious, some political. Almanacs were especially popular, not simply for their calendars with sunrise and sunset times, moon phases, and weather forecasts (all adjusted for local conditions), but also for their currency and measurement tables, their lists of government officials, their agricultural and domestic advice and other practical information. Some almanac

Above: Spacers that sit lower than the type separate words and fill in to justify lines to the desired width. Left: Set lines of types fill a galley in preparation for proofing.

editors gave their publications political colour by inserting quotes and commentary.

Thanks to both immigration and education, an increasing number of Canadians were literate. Schools were organized with the primary purpose of teaching children to read. Families gathered at firesides where one member read aloud to others. In village and crossroad taverns, customers gathered to hear newspapers read aloud and to discuss their contents vigorously.

To the first generations of immigrants, whether from Britain or the United States,

A proofing press prints a galley, which a proofreader checks for typographical errors and marks for correction.

newspapers provided an essential link to the world beyond their immediate neighbourhood. Along with local ads and public notices, newspapers contained news from back home — intelligence of wars and rumours of wars, announcements of important births and deaths, reports of fires and shipwrecks, details of political shifts and social controversies, lists of market prices and customs duties — all reminders that the wilderness of Canada was very much connected to a larger world. The papers also offered readers recipes and reme-

A proofing press in action.

dies, amusing anecdotes and sentimental verses, philosophical reflections and editorial commentaries.

There was a growing market for political debate. When British administrators brought printing to Canada, they never intended the press to be free. The first King's Printers, as we have seen, were dependent for their livelihood on the goodwill of colonial officials. Independent printer-editors, however, had little to fear from government regulation. Along with serving their local customers, some of them assumed a larger role, using their columns to voice opinions, challenge policies, expose errors — and even promote candidates at election time.

In English-speaking Canada, in the 1820s and 1830s, two editors in particular began to attract notice and exercise influence: in Upper Canada, William Lyon Mackenzie; in Nova Scotia, Joseph

A printed galley proof can now be proofread.

Howe. Though different in background and temperament, each would profoundly affect the political development of his province and, indeed, of the emerging nation.

In 1824 Mackenzie, a Scot with four years' profitable experience as a merchant in Upper Canada, launched in Queenston his *Colonial Advocate,* the first independent paper in the province to have significant political impact. Its intent, as its name suggested, was to speak for the farmer-settlers, not for the government. In its ten-year existence, its editor never stopped attacking a colonial administration he saw as incompetent, ineffective, and worse,

A galley of set types is corrected after proofreading.

expensive. An editorial in the first issue derided Lieutenant-Governor Peregrine Maitland: "[Despite] his enjoyment of a princely salary ... we cannot remember any thing he has done of a public nature worth recording."

A few weeks later Mackenzie boasted: "To become an advocate for the rights of this people, is to become an enemy of the ruling power." The target of his continued assault was the ruling elite — the "Family Compact," he would later dub them. An election was in the offing and Mackenzie urged his readers to assert themselves: "Up then, and be doing. Stir yourselves ... and strike at the roots of corruption, in the person of our late corrupt representatives." After six months in Queenston, he abandoned the role of an editor of a small-town weekly and moved the *Advocate* to York, the centre of political action.

Hundreds of copies were printed each

Cheap, portable, and durable, the Washington press was, in its time, the most widely used press in North America.

week. In Upper Canada, nearly everyone was reading the *Advocate*, but not everyone was paying for it — subscribers and advertisers were often delinquent. The most widely circulated paper in the province was anything but profitable. Indeed, but for the excessive zeal of its opponents, the paper would have failed. A gang of men, all associated with the ruling establishment, decided to silence the fiery editor. They broke into his printing office in broad daylight, smashed his press, scattered equipment, and threw cases of types into the bay. The destruction was the salvation of Mackenzie. He sued his assailants, won his case, and collected enough in damages to pay off debts, make repairs to his press, and start up again, now hailed as a hero the government could not suppress. A few weeks before the "Types Riot," as it was dubbed, the outspoken editor had written:

> *[T]he true friend to his country ... will always be ready at the moment of danger to sound the alarm, to arouse the sleepers, and to awaken the careless. Riches and preferment he counts as dross — his desires are few — his wants easily supplied — his country is his and what would he more!*

A brayer is used to ink the types in a forme.

Two years later, Mackenzie was elected to the House of Assembly where he became a vigorous advocate of reform and critic of government waste and excess. He toured the province, organizing meetings and demanding change. In 1834 he was elected Toronto's first mayor. In spite of all this political activity, he continued his work as a journalist, first with his *Advocate* and later with a new paper, *The Constitution.* Ultimately, in 1837, out of office and frustrated with electoral politics, he fomented an ill-fated rebellion — when it failed, he fled to the United States with a price on his head. It was more than a decade before a general amnesty enabled him to return to Toronto, but while in the United States, and later back in Upper Canada, he never stopped writing.

Joseph Howe was fourteen when he began his apprenticeship in his father's printing

business in Halifax in 1818. John Howe was an ardent Tory — and the King's Printer — publishing both the *Royal Gazette* and the *Halifax Journal*. In 1827 Joseph, not yet twenty-three, took over another Halifax paper, the *Weekly Chronicle*, re-christening it the *Acadian*, a journal devoted to industry and letters with little hint of anything but the most loyal brand of politics. A year later he purchased the *Novascotian*, the second most circulated paper in the province, and affirmed his attachment to the Empire by adopting as the paper's motto "The Constitution, the *whole* Constitution, and *nothing but* the Constitution."

Over the next few years, Joseph Howe became an increasingly ardent Nova Scotian,

Top: Turning a crank moves the bed into the body of the press.
Above: Images of Franklin and Washington adorn a Washington press.

Types were picked one at a time to compose a line.

and in the process became increasingly frustrated with the British administration of the colony and its appointed Council:

> *We have here twelve men who can jeopardize the peace, and destroy the Revenue of the Country; who can … inflict evils of a nature so multifarious and extensive as to require the passage of years to repair; and yet, can turn round and shift the responsibility from their own shoulders.*

Like Mackenzie, though initially with far less fervour, Howe became a champion of "responsible government," advocating an elected provincial government, under the auspices of the British Crown but essentially accountable to the voters of Nova Scotia.

The success of the *Novascotian* carried Howe into the provincial parliament and, eventually, away from printing. In 1835, while still an editor, he was prosecuted for criminal libel for publishing a letter censuring the local magistrates. Tried before a jury, he defended himself by defending

Compositing required dexterity and attention to detail.

his newspaper and freedom of the press:

> *I conjure you to judge me by the principle of English law, and to leave an unshackled press as a legacy to your children. You remember the press in your hours of conviviality and mirth — oh! do not forget it in this its day of trial.*

He may have been guilty according to the law, but the jury readily acquitted him. As the Types Riot did for Mackenzie in Upper Canada, this libel trial made Howe a popular hero in Nova Scotia. For the next forty years Howe would be a central public figure in the province, in imperial relations, and in the first years of the Dominion.

Though politics was their passion, both Mackenzie and Howe earned their livings by printing. Howe learned his trade as an apprentice to his father. Mackenzie managed a printing establishment and called himself a printer, but he lacked the dexterity to set types or pull presses: "I am awkward at dissecting a still jointed turkey ... awkward at mending a pen ... awkward at chopping firewood, or mowing hay."

The growing importance of independent printer-editors was partly a product of improvements in printing technology. About 1800 wooden common presses began to be replaced by the wood-and-iron fabrications of Adam Ramage of Philadelphia. "Ramages" quickly spread everywhere, but in the mid-1820s iron presses took over the industry. They were cheaper and easier to maintain. More important, they were more durable and faster, requiring only a single pull by a pressman. Although there were various competing models (see Appendices), the "Washington" press (so called for its relief decoration of George Washington) became the workhorse flat-bed press in printing offices throughout North America.

With continued use and re-use, a printer's types became worn and required replacement. The first Canadian printers bought their types from American foundries. Mackenzie purchased his supplies and equipment from New York — and complained about excessive import duties. By the 1830s a type foundry was operating in Montreal.

About the beginning of the nineteenth century there was a shift from Caslon "old style" to "new style" types. Among other changes, new style did away with the long *s*, which could be confused with *f*. Caslon's basic designs continued to dominate, but many printers augmented Caslon with other type styles, such as Egyptian, Tuscan, and Gothic. Thanks to new methods of manufacture, wooden letters of different sizes and varying designs became more widely available for printing headlines and posters.

Stereotyping, printing from a solid metal plate cast from a forme of set types, was developed in Europe late in the eighteenth century. Stereotype plates could be kept for future printings of a work, while the original types were freed for distribution and re-use. Few Canadian printers cast their own stereotypes during this period, but some purchased them to produce their own editions of works previously published elsewhere. In the 1820s, Mackenzie warned readers that an American publisher was misrepresenting an edition of the Bible he was peddling as the product of a Canadian printer when in fact it was printed in the United States using American-cast stereotypes. (A decade later, Mackenzie himself bought American stereotypes to produce his own edition of the Bible.)

Until mid-century, all paper, even paper for newspapers, was made from linen or cotton rags. Canadian paper mills started operation in 1804 in Quebec and 1819 in Nova Scotia. At Mackenzie's urging, in 1826 the legislature of Upper Canada offered a reward of £125 to the first paper mill to start up in the province

— within months two mills were in operation. Since paper mills were chronically in need of rags, print shops collected them from their customers as payment for their newspapers.

Regardless of their political proclivities, printer-editors across the country faced similar demands. Every week there was a newspaper to put out — a four-page broadsheet to be filled with news and commentary, helpful advice, and homely amusements. Subscribers accounted for much of the income, but advertisements paid the bills — at least one full page of paid ads was required for an issue to begin to be profitable.

The remaining columns of the paper were filled with words — ten to fifteen thousand words a week. One source of copy was other newspapers. (Mackenzie claimed he read a hundred papers a week!) Editors — in town, in the province, in Canada and the U.S., even across the Atlantic — regularly "exchanged" papers looking for news, ideas, and "filler": poems, anecdotes, household hints, and other amusing, educational, or inspirational material. Some copy — London newspaper reports of British Parliamentary debates, for example — would be lifted wholesale. Other items were summarized as brief squibs, simply titled "Foreign News" or "United States." Occasionally editors would insert extracts from the pamphlets they were printing using types already set.

Print shops collected linen and cotton rags on behalf of the paper mills.

A newspaper also provided a personal forum for the editor to report events he witnessed or interviews he conducted. His political and cultural opinions defined the paper. Sometimes his commentary was prompted by something a competing editor wrote, or failed to write. Cross-paper debates were common and might continue back and forth for weeks, which may have been confusing for readers who only took

Some subscribers picked up their newspaper at the post office.

Country post offices were also general stores selling a variety of goods.

one of the papers involved. One way or another, every week all the columns of a paper had to be filled.

A skilled compositor took at least two full days to compose, proof, and impose the copy for a week's paper. When pages one and four were locked in formes, printing the first side of the paper could begin. It took at least three hours to print five hundred copies (Mackenzie circulated twice that many). The next day pages two and three, with the most up-to-date news and commentary, went to press. On the third day papers were folded, rolled and tied, hand addressed, and taken to the post office. Papers also might be privately delivered locally or picked up at a news agent or the printshop itself.

Subscribers and advertisers had to be

constantly reminded to pay their bills. Howe regularly made long tours of the province — to scout the political landscape, but mainly to promote his publications and collect money from delinquent subscribers. During his weeks away, his wife managed the business while Howe kept in touch by mail:

Some printer-editors included bookbinding as part of their operations.

> *My dear Susan Ann … I enclose two Orders … making £18. I also sent you £10 in dollars, which you can sell for whatever premium dollars are bringing … I send likewise £2 in gold.… Do not neglect your dunning — the more money you can collect the better.… May God bless you till I return and help you to get through all the cares which have been laid upon you by*
> *Yours affectionately, Jos. Howe.*

An independent printer-editor had bills to pay, debts to collect, correspondence to keep up, supplies to order, equipment to maintain, rent to pay, and payrolls to meet. Journeymen printers were paid weekly — in this period £7 or £8. They were notoriously transient, moving from shop to shop, quitting or being fired, skilled in their work but often delinquent in their personal lives. And adolescent apprentices required constant supervision.

Perhaps it is no wonder Mackenzie was prompted to muse:

> *The proprietor and editor is and must be a slave. He toils from day to day to get forward; he perceives that his ideal profits are accumulating; that the list of his debtors is rapidly augmenting; that his prospects are slowly brightening; but he still must remain a slave.*

3 The Voice of the People

The Community Printers, 1850–1900

During the second half of the nineteenth century, the business of printing, like the country itself, was moving in two different directions. As cities such as Montreal and Toronto became more diverse and industrialized, their printers diversified and specialized. As new territories opened to settlement, other printers packed up their types and presses and moved west, creating small-town print shops not unlike those of the

Print shops opened in small towns during the second half of the nineteenth century.

Printer-editors were both craftsmen and businessmen.

Country print shops also sold books and stationery.

printer-editors of earlier generations.

In the growing urban areas, printers began serving different markets. While some continued to run small businesses doing short-run job printing, others began to concentrate on publishing books — text books, law and medical books, religious books, and children's books. Other printers produced huge runs of catalogues or promotional matter. Still others found a market for fine printing and moved from letterpress printing into engraving or lithography. But most prominent, perhaps, were the printers who became publishers of mass-circulation daily newspapers.

The increasing demand for quantity and speed stimulated continual innovation in printing technology. Despite attempts to convert hand presses to steam, real change came with the invention of the cylinder press. Different designs all employed the same basic principle: paper attached to a large heavy drum or cylinder and rolled over an inked forme on the flat bed of the press. A hand-operated cylinder press could print 1250 sheets an hour — five times the output of a Washington. When harnessed to steam power, cylinder presses printed thousands of sheets an hour. Later models made use of several cylinders and as many as ten operators working together. Newspaper publishing depended

The cylinder press, developed in mid-century, could print much faster than flat bed presses.

on the capacity to print many copies very quickly — rolling cylinders literally revolutionized the industry.

Newspapers were not only adopting new technologies, they were evolving ideologically. Official *Gazettes* became specialized publications for printing legal notices. Newspaper editors were largely free from direct control, but government exercised its influence in other ways — privately persuading editorial policies and selectively placing paid advertising in sympathetic papers. Political journalism evolved into papers that were pointedly partisan, often subsidized by private sponsors. And more papers appeared that appealed to religious, trade, or other specialized groups of readers.

One man who eagerly adapted to this brave new world was George Brown. Born in Scotland, he had emigrated to New York and then to Toronto to start, with his father Peter Brown, a Free Church Presbyterian paper, the *Banner.* George's interests soon became less

Paper is attached to a cylinder and rolled over an inked forme of set types.

parochial and the *Banner* allied itself with the Reform Party. The next year, 1844, George Brown launched the *Globe*. As its name suggested, its outlook encompassed the province, the British Empire, and the world. Announcing his new publication, Brown affirmed his faith in political parties:

> *It is usual … [for a new newspaper] to declare against all political partisanship … [we, however, are] convinced that no strong Government has ever existed — no great measure has ever been carried — without a combination of persons holding similar views. THE GLOBE will therefore strenuously support the party which shall advocate the measures believed best for the country.*

In the course of the next decade, the weekly *Globe* became a bi-weekly, a tri-weekly, and

The speed of the cylinder press made mass-circulation daily newspapers possible.

in 1853, a daily paper in competition with four other dailies. Brown proceeded to buy out some competitors, and drive out others — in time, the paper's circulation was more than 20,000 for the daily edition, plus nearly 30,000 for the weekly edition that served the province and the country. By 1860 it was Canada's largest newspaper.

Like politically minded editors before him, in the 1850s Brown eagerly entered politics. He quickly became the Reform Party leader; his principal foe was Tory leader John A. Macdonald. In the Parliament of Canada (Lower and Upper Canada were conjoined in 1841), these two men bitterly opposed each other. Reform and Conservative rivalries intensified, aggravated by tensions between French and English, and Catholic and

Protestant, frustrating effective government for nearly two decades. At last, Brown softened his opposition and reached an accord with Macdonald that led to Confederation and the founding of the Dominion of Canada. That year Brown resigned from Parliament, but he continued to exercise his considerable influence through the columns of the *Globe*. He died in 1880, shot by a disgruntled employee.

Throughout much of his career as a publisher, Brown had difficulties with his labour force. Among the first unions in Canada were those organized by journeymen printers. The York Typographical Society was formed in 1832 — Mackenzie's son James was an officer. Disrupted by the Rebellion of 1837, it was revived a decade later as the Toronto Typographical Union. In large urban printing establishments unions were coming into their own — inspired in part by the effective collective action of American printers. Their major demands were shorter hours, fewer (essentially unpaid) apprentices, safer environments, and of course, better pay. Throughout the second half of the nineteenth century (and much of the twentieth) labour strife among the typographical workers was notorious, not only in Canada but through much of the industrial world.

Another controversial issue was the presence of women in the industry. From the beginning of printing in Canada there were women, albeit in small numbers, working in printing: Bushell's daughter set types in her father's shop, and occasionally operated the press; Howe's wife published the *Novascotian* when he was out of town; Mackenzie sometimes recruited his daughters to sew pamphlets.

Although a few might learn printing skills in a small family-run print shop, no girls were formally apprenticed to become qualified journeymen printers. But some employers did train women to be compositors — they learned quickly and would work for less money, much to the dismay of unionized male journeymen. George Brown's employing of women — "Brown's harem" male journeymen derisively dubbed his establishment — was a major reason for the first strike at the *Globe*. The unionized journeymen won that round; the women were let go. But the question of women working in print shops or joining unions continued to be a bone of contention in the industry, both in Canada and the United States, until well into the twentieth century.

There were also several examples of women (in most cases succeeding their deceased husbands) running newspapers in Upper Canada. One noted woman publisher was Mary Ann Shadd, an immigrant from the United States who promoted Canada as a haven for African Americans, both slaves and freemen. Demonized by Henry Bibb, editor of the *Voice of the Fugitive*, as a woman whose "duplicity is sufficient to prove a genealogical descent from the serpent that beguiled mother Eve," Shadd's launch in 1854 of the *Provincial Freeman* made her the first African American woman to publish and edit a newspaper in North America. Despite economic difficulties and the vicious opposition of other editors — all male, some black — the paper persisted for

Women occasionally worked in country print shops — in urban shops their employment as compositors was opposed by unionized journeymen printers.

several years, championing black nationalism and women's rights. And its voice was distinctly Canadian:

> *None of the papers published by our people, in the States, answer our purpose. They either pass us by, in cold contempt, ignore us altogether, keep themselves and their readers, or both, ignorant of what Canada is, or in some other way, by opposition or neglect disparage us, as much as convenient.*

From the earliest years of printing, the most labour-intensive aspect of the process was composing — hand-picking types, justifying lines as they were set and, after they were used in printing, distributing types back into the appropriate cases. Although there had

Women were not formally apprenticed, but some learned the skills informally and became printers.

been many attempts to mechanize the process, it was nearly the end of the nineteenth century before Ottmar Morganthaler invented a successful typesetting machine, the linotype. With it an operator at a keyboard could mechanically combine the matrices for individual letters to cast a full line of types as a single lead slug. After use, slugs were melted down for recasting, thus eliminating the task of distributing used types.

Competing, but somewhat similar line casting machines — the Rogers Typograph and the Monoline — were also produced. Barred from the American market by

Compositing was the most labour-intensive part of the printing process.

Morganthaler's patent rights, their cheaper price and simplicity of operation made them popular in small Canadian newspaper offices. The Monotype typesetter — which employed a keyboard to create a paper tape that was then fed through a separate casting machine — was slower than a linotype, but its flexibility made it especially useful for book publishing. Also, the Ludlow Typograph was introduced — the many sizes and designs of its typefonts simplified casting headlines and display ads.

Printing became even more efficient with the advent of new ways to make stereotypes. Papier mâché molds made on flat formes of set types were curved to cast curved stereotype plates. The process made possible a new kind of press, the rotary press, with the stereotype of a page mounted on a cylinder. As the cylinder rotated, its typeface was inked and pressed onto paper fed through the press not as sheets but from a great roll or web — hence "web presses," as they were called. Multi-cylinder web presses

Community printers were jacks-of-all-trades.

could print many pages almost simultaneously at high speeds — tens of thousands of copies of whole issues of a paper an hour.

As the century progressed, newspapers, magazines, and books increasingly added images to their texts, especially after engraving was combined with photography to produce photoengraving. Pictures could be printed in shades of grey — or, using different plates and inks, even in colour.

These changes at large printing establishments were matched by improvements in smaller scale printing. The need for faster, cheaper, and more precise ways to print cards, tickets, and business forms spurred the development of the job platen press. Its platen held

Turning a key moves metal quoins to lock types set in a forme.

a forme of set types at an angle rather than flat. Opposite and also at an angle, its tympan held a sheet of paper. As an operator pumped a foot treadle, one continuous mechanical action rolled ink over the forme, brought the tympan and platen together, separated the two to allow the printed sheet to be replaced with a blank, and repeated the process — a thousand or more cycles an hour. The most popular jobber, the "Gordon" (first introduced by George P. Gordon in 1851), was manufactured in several sizes — larger models could be powered by steam engines or, later, electric motors.

Hand-operated flat-bed presses were not yet obsolete, however. Despite their more moderate speed, they continued to be employed for proofing or for short runs of posters or handbills. Or they were sold by urban printers to their smaller country cousins who continued to print in the old ways.

These men, unlike the specialized printers in the city, were jacks-of-nearly-all-trades — news-gatherers, writers, editors, compositors,

For many centuries, type surfaces were inked using hair-filled, leather-covered, wooden-handled balls.

pressmen, mechanics, accountants, bill-collectors, publicists — and civic leaders as well. Commercial customers wanted letterheads and cards; individuals asked for invitations or funeral notices; local businesses ordered circulars to publicize special sales and auctions; promoters used posters to announce fairs and special events; and politicians needed signs and handbills to court voters during elections. But a nineteenth-century country print shop's most important publication was a weekly newspaper. Its columns of ads, notices, news, commentary, advice, poetry, and trivia provided a common point of reference for everyone in a town.

The personality of the printer-editor determined the character of the paper. The editor's opinions — on local issues, on municipal elections, on provincial and national politics — might not be shared by all citizens, but everyone read them. Some editors were

Above: The platen press's fly-wheel operation requires placing and removing sheets at a steady pace as the press opens and closes.
Top left: The inking mechanism of a platen job press.

frankly partisan, declaring their firm allegiance to the Tories or Liberals, even running for parliament. Others were like Henry Ball — editor of the *Advertiser*, the first paper in Creemore, Ontario — who proudly boasted his broadsheet was "Independent in Everything, Neutral in Nothing."

A country print shop was a comfortable place compared to the large urban establishments. Large windows let in light for setting types, reading proofs, taking notes, and keeping accounts. The press area was kept heated and well-ventilated to allow ink to dry and prevent paper from growing mold. The cluttered shop was the site of constant human activity, as customers came and went, collecting their newspapers, placing orders for job

Even after the invention of mechanical typesetting, large wooden types were hand set to make headlines.

The voice of each printer-editor was unique, reflected in the articles and opinions of the newspaper.

printing, or making purchases from the stocks of paper, pens, ink, sealing wax, notebooks, and stationery supplies that the printer offered for sale. He also sold forms and pamphlets printed on the premises and acted as sales agent for magazines and newspapers published elsewhere. Most country printers were also booksellers, offering the works of Canadian, American, and British publishers.

These country printing establishments were much more than commercial offices. Print shops did more than print. They were places where people met to hear the news, share gossip, and exchange ideas. They were focal points that provided information to strangers and introduced immigrants to their neighbours. They helped make towns and villages genuine communities — especially, perhaps,

An operator places and removes sheets at a job platen press.

the newer settlements that were opening in the western half of the nation.

In 1859 William Buckingham and William Caldwell hauled a second-hand Washington press from St. Paul, Minnesota, to Fort Garry (Winnipeg) to commence publication of the *Nor'Wester*, the first newspaper on the prairies. In a few years both men abandoned the west for eastern comforts, but their printing enterprise struggled on under new management. In 1869, Louis Riel and his Metis followers occupied the *Nor'Wester* print shop and printed a broadside addressed "To the Inhabitants of Rupert's Land" on the well-worn Washington. Riel's printed words were issued to mark the outbreak of his first uprising. According to historian Eli Maclaren, Riel's having access to a press to publicize his message in 1869, but not in 1885, constituted "a crucial difference between the diplomatic Red River Resistance of 1869–70 and the destructive Northwest Rebellion of 1885."

Further west, in Victoria, in 1856, Roman Catholic Bishop Modeste Demers imported an aged French hand press intending to publish religious materials. It sat idle for two years until an American printer, Frederick Marriott, used it to produce the short-lived *Vancouver Island Gazette* to compete with British Columbia's first newspaper, the *Victoria Gazette.* Marriott next used the press to print two newspapers simultaneously, one in French, the other in English — neither succeeded. Later that same year, the "Demers press," as it is called today, was printing a fourth newspaper — one that lasted much longer and exercised much greater influence, the *British Colonist.*

The *Colonist*'s founder was Amor De Cosmos. Born William Alexander Smith in Nova Scotia, De Cosmos legally assumed his extraordinary new name to express his "love of order, beauty, the world, the universe." He

Printing offices were open, warm, and well-ventilated, often making them pleasant places to work.

worked for a number of years at various jobs in Californian gold mining towns until the Fraser River gold rush of 1858 lured him north to Victoria. Once there he quickly resolved to offer his opinions via the columns of a public journal and launched the *Colonist*. In his first editorial he asserted:

> *We intend, with the help of a generous public, to make the* British Colonist *an independent paper, the organ of no clique or party — a true index of public opinion.*

Printed broadside posters informed the public about upcoming social and political events.

With words that might have been written three decades earlier by Joseph Howe, De Cosmos continued:

We shall ever foster that loyalty which is due to the parent government, and determinedly oppose every influence tending to undermine or subvert the existing connection between the colonies and the mother country.

Yet, with sentiments echoing Mackenzie's, a few weeks later De Cosmos took aim at the colony's governor, James Douglas:

> *He alone is responsible for all the corruption, peculation, wrongs, outrages, depopulation, obstructions, mistakes, and losses, which have been entailed on these colonies through corrupt and unfit officials.*

Like many of the printer-editors before him, De Cosmos entered politics, assuming leadership of those urging political reform. He advocated uniting Vancouver Island with the mainland to form a single colony with responsible government. Later, to prevent its annexation by the United States, he championed the cause of British Columbia's joining the new Dominion of Canada. Elected to represent Victoria in the House of Commons in Ottawa and, from 1872 to 1874, simultaneously serving as British Columbia's second premier, he lobbied successfully for the completion of the national railway and unsuccessfully for its extension to Victoria. He also published another newspaper, the *Daily Standard.* Despite his editorial influence and his political eloquence, he was eccentric and controversial, making many more enemies than friends. Amidst accusations of scandal involving mining investments, De Cosmos abandoned electoral politics and retired to private life. He died in 1897 after fifteen years of decline ending in total mental breakdown. Few persons attended his funeral or paid much attention to his death. As a letter in the *British Colonist,* the paper De Cosmos founded, aptly observed: "Governments, corporations and

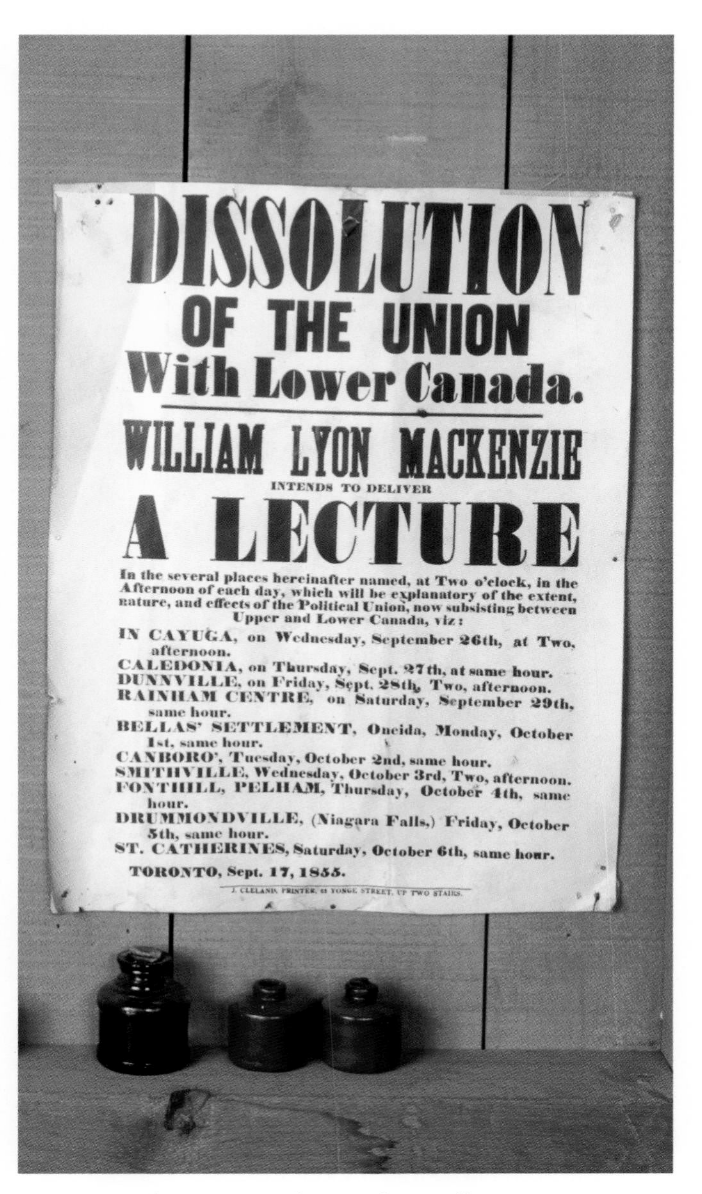

A poster advertising a lecture by William Lyon Mackenzie.

Weekly newspapers connected members of a community to each other, and to the larger world.

the public seem to have no hearts, no sentiment, no memory — callous to all but their own interests and affairs."

The Demers press went on to print the *Cariboo Sentinel* in the 1870s, produce job work in Kamloops in the 1880s, and finally, in 1908, it reclaimed its original Roman Catholic identity. It was acquired by the Sisters of St. Ann, an order first brought to British Columbia by Bishop Demers. The press is now on display at the Royal British Columbia Museum in Victoria.

Postscript

Letterpress printing continued to flourish in Canada well into the twentieth century, especially in large printing establishments and daily newspapers. Gutenberg's basic process — inked metal types pressed to paper — persisted for centuries, with advances in technology bringing increased speed, labour efficiency, and improved print quality. During the middle years of the century, the whole industry underwent a revolution, adopting

Although not used commercially anymore, individual wood and metal types are still used by artisans today.

Old equipment was eventually sold to artisans or abandoned.

Printed broadside posters.

offset printing and photo typesetting, "cold type" replacing the older lead-cast "hot type." Further adaptations and inventions, especially the advent of computers and high-speed data transmission, totally doomed letterpress printing as a commercial enterprise.

For many decades, country newspapers were produced in the old way, augmented by electrically powered presses and typesetting machines as these could be afforded. In time, however, competition forced these print shops to abandon letterpress. Much of the old equipment was junked as scrap metal or sold to private artisans.

Letterpress printing may be obsolete, but it

Freedom of the press remains an important principle today.

lives on unrecognized, embedded in our language. We all know the difference between upper and lower case letters. Printers and editors still speak of typesetting copy, changing fonts, inserting cut lines, leading lines, or laying out pages. We tell our children to mind their p's and q's or ask the long-winded to cut to the chase — and who among us does not sometimes feel out of sorts?

Today typography is an aesthetic and historic rather than an economic endeavour. Hand-printed books and other works continue to be produced by hobbyists and skilled artisans. Students at colleges of art and design employ old types and hand presses in their creative endeavours. And — as the photographs in this book amply indicate — Canadian museums and historic sites have extensive collections of letterpress equipment and feature printing in their public displays and demonstrations.

Freedom of the press is a central principle of our Canadian culture, affirmed in our Charter of Rights and Freedoms. It was not always so, as the stories of printers in this volume testify. In the first Canadian newspapers, King's Printers published the voice of the government speaking to the people. In time independent printer-editors acted as the voice of the people speaking to the government. Gradually, especially with the spread of community newspapers, the press assumed a third role, giving voices to people speaking to each other.

In our time, these voices find many different forms of expression. The growth of electronic media has almost eclipsed printing on paper. Newspaper circulation has declined and far fewer community newspapers exist. Yet the need for news and information in print form has never been greater. And more, not fewer, books are being published than at any other time in our history. Compared to the era described in these pages, the press today may be less powerful, but it remains a potent force.

Appendices

Historically Significant Hand Presses

Based on:

Stephen O. Saxe, *American Iron Hand Presses.*

David S. Rose, *Introduction to Letterpress Printing in the 21st Century,*

Information provided by printing historian Stephen Sword.

Five Roses Press, www.fiveroses.org/intro.htm

Briar Press Museum, www.briarpress.org

Wooden Presses

Adapted by Gutenberg from wine-making presses in use long before his time, their design remained virtually unchanged until late in the eighteenth century. These presses employed a large screw mechanism with a long lever to make the impression. Also known as screw presses or common presses, most North American wooden hand presses were imported from England. Easy to transport and repair, they continued to be used through much of the nineteenth century even after more advanced presses became available. They were usually operated by a pressman, who handled the paper and pulled the impression, and an assistant, who applied the ink. Although estimates vary, later models of these presses could print up to 250 sheets per hour.

Roy Press

Constructed in Quebec sometime in the eighteenth century, this common press was used by Louis Roy as the first press in Upper Canada. It is now on display at the Mackenzie Printery and Newspaper Museum in Queenston, Ontario

Ramage Press

A number of wooden-framed presses were constructed by Adam Ramage in Philadelphia in the early decades of the nineteenth century. They incorporated an iron platen and an iron screw mechanism for raising and lowering the platen, designed, according to Ramage, "to render the hand press efficient, simple in construction, and moderate in price." These durable presses were especially popular with smaller newspaper offices.

Iron Presses

At the beginning of the nineteenth century, presses made almost entirely of iron began to appear. Most of these presses also made use of toggle mechanisms, an improvement over the traditional screw. Two operators working together could print large sheets — two newspaper pages — at a rate of 300 per hour,

Stanhope Press

The first all-iron printing press was developed in England in 1803 by Charles, third Earl Stanhope. The press used a screw mechanism along with complex levers to increase its power. A few were imported to North America.

Columbian Press

Invented by George Clymer in Philadelphia in 1813, it used compound levers which increased the force of the impression while demanding of the printer relatively little effort. A cast-iron eagle atop the press acted as a counterweight to raise the platen after an impression. A number were sold in North America, but Clymer achieved greater success in England.

Wells Press

Invented by John Wells in Hartford, Connecticut in 1819, it replaced the screw with a toggle mechanism to apply downward pressure on the platen. Later models used springs instead of counterweights to ease raising the platen after each impression. Joseph Howe purchased a Wells press in 1829.

Smith Press

Introduced by Peter Smith in 1821 and manufactured by Robert Hoe & Co. in New York, the press used a toggle mechanism mounted in a massive cast iron acorn frame. Despite competition from the more popular Washington presses, Smith presses continued to be manufactured throughout much of the nineteenth century. William Lyon Mackenzie purchased a Smith press in 1825.

Albion Press

Invented in London by Richard Whittaker Cope about 1822, it continued to be manufactured throughout the nineteenth century. Its spring and lever mechanism was more compact than that of other iron presses. Very popular in England, a number also found their way to Canada for the finer printing of books and other documents.

Washington Press

This press was developed over a period of years by inventor Samuel Rust in New York City. His first models introduced an improved toggle mechanism that offered greater leverage than those of the Wells and Smith presses. By 1829, Rust had developed a readily dismantled frame in which wrought-iron rods ran inside hollow cast-iron columns, making the press lighter and easier to transport, a major factor in its commercial success. After Rust's patents were acquired by R. Hoe & Co. in the 1830s, several thousand Washingtons were sold in North America. The press was also copied by other manufacturers, including Hall Manufacturing in Oshawa and Westman & Baker in Toronto.

Tufts Press

Invented by Otis Tufts of Boston in 1831, it combined the acorn frame of the Smith press with a toggle joint similar to the Wells press. For many years the presses were manufactured in Boston by Isaac Adams, an early developer of power presses. Although not widely used in Canada, Tufts presses are found in the collections of the New Brunswick Museum in Saint John and Black Creek Pioneer Village in Toronto.

Cylinder Presses

First invented in Germany by Friedrich Koenig in 1814, flatbed cylinder presses did not come into general use in North America until mid-century. Among the most popular were those made by R. Hoe & Co., which were especially useful for large sheet book and newspaper publishing and could print 800 to 1500 sheets per hour. Later adaptations of the drum cylinder principle led to cylinder or rotary presses being adopted for most large-format, high-volume printing applications. Hand-cranked presses required two operators, one to crank the mechanism, one to feed blank sheets. Powered cylinder presses, in spite of their size and speed, needed only one operator whose main task was handling the paper. In their heyday, cylinders carried a huge part of the workload of a printshop.

Platen Job Presses

Beginning with Stephen P. Ruggles of Boston in the 1840s, these presses became the basic design for nearly every letterpress to follow. Often called "clamshell" presses, the platen and bed were hinged below their lower edges to close on each other with each impression. They were operated by one person working a foot treadle and later with an engine drive. Initially designed for short-run job work, they were much faster than hand presses and were especially useful for small sheet work. Their size, weight and ease of operation led to their adoption by many community print shops. By the end of the nineteenth century, more than a hundred different models were on the market, known by varying names: platen presses, jobbers, trade presses, floor-model presses, or rotary presses.

Gordon Franklin Press

Introduced by George P. Gordon in 1851, this popular press set a standard for platen job presses. As the various patents expired, many entrepreneurs began to produce their own "Gordons."

Chandler & Price Presses

In 1886 in Cleveland, Ohio, William T. Price and Harrison T. Chandler began production of presses based on the Gordon Franklin. More than forty thousand "C & P oldstyle" presses, featuring lovely curved spokes on their flywheels, were manufactured until they were replaced in 1912 by "new style" presses with straight flywheel spokes and heavier frames.

Lever Presses

These presses, made by a variety of companies in a number of designs, were scaled-down versions of platen job presses, employing a large lever handle to bring the bed and platen together for an impression. Small enough to be placed on a table or cabinet, they were also known as tabletop presses, card presses, amateur presses, or toy presses.

Excelsior Press

William A. Kelsey began making inexpensive presses for amateurs in 1872. His Excelsior press became so popular that it prompted many competitors.

Galley Proof Presses

Also called roller proof presses, they were used to make proofs of galleys of type. Their simplicity of operation — rolling a small heavy cylinder over a galley to make an printed impression — meant that these presses could also be used for high quality, limited-edition print jobs.

Ink and Inking

Printer's ink was a thick black paste. Early printers made their own ink from lampblack (fine carbon powder) and linseed oil. In the early decades of the nineteenth century, manufactured ink shipped in kegs became available and was much easier to use. For a long period, printing in colours other than black was virtually unknown, in part because of the extensive cleaning required to change from one colour to another.

William Lyon Mackenzie once published notes, extracted from a collection of 500 recipes, on making two colours: "*Printer's Red Ink* — Soft varnish and vermilion with white of eggs, not very thick. Common varnish, red lead and orange. ... *Blue Ink for Printers* — Prussian Blue and a very little Ivory black with varnish and eggs, very thick. Common Indigo and varnish; then wash off with boiling lees."

For many centuries, inking a forme was done with two ink balls — handmade, leather-covered, stuffed with wool, and wooden handled. Ink was transferred from ink stone to type by rapidly, *ratta-tat-tat* pounding the balls on the type. Printers' balls were difficult to make, hard to clean, and jealously protected by their owners from use by other printers.

By 1830, roller brayers replaced ink balls for inking. The surface of a brayer was coated by rolling it on an ink stone and then rolling it over the type in a forme. For many years these roller brayers were "composition," that is, a mixture of horse glue and molasses, glycerine and other chemicals (exact formulas were kept secret). Brayers cracked easily and became pitted; in storage, they could deform and grow mold. Printers needed special equipment to reshape them and restore their surfaces. The introduction of rubber rollers was a significant improvement — they were commercially made, lasted a long time, and were easier to use and to clean. Inking by rolling was a crucial step to successfully employing steam power printing presses.

Sites of Interest

For museums from which information was obtainable, the press or presses and other relevant equipment in their collections are indicated.

* Sites marked with an asterisk offer demonstrations of their letterpress equipment.

Newfoundland

Road To Yesterday Museum
Water Street
Bay Roberts, NL AOA 1GO
Phone: 709-786-3482
ejerrett@nf.sympatico.ca

Gordon Franklin press, platen job press, linotype

*Sir William F. Coaker Heritage Foundation
Main Street
Port Union, NL A0C 2J0
Phone: 709-469-2207
coaker.foundn@nf.aibn.com

platen job press

Trinity Museum
Church Road
Trinity, NL A0C 2S0
Phone: 709-464-3599
ttci@nf.sympatico.ca

lever press

New Brunswick

*Kings Landing Historic Settlement
20 Kings Landing Road
Kings Landing, NB E6K 3W3
Phone: 506-363-4999
www.kingslanding.nb.ca

platen job presses (3)

New Brunswick Museum
Market Square,
Saint John, NB E2L 4Z6
Phone: 506-643-2300
nbmuseum@nb.aibn.com

Tufts press

*Village Historique Acadien
14311 Route 11
Caraquet, NB E1W 1B7
Phone: 506-726-2600 or 1-877-721-2200
www.villagehistoriqueacadien.com

Washington press, platen job press

Nova Scotia

*Acadian Museum and Archives of West Pubnico
#898 Autoroute 335, via Route 3
West Pubnico, Yarmouth, NS B0W 3S0
Phone: 902-762-3380
musee.acadien@ns.sympatico.ca

platen job press

*Nova Scotia Museum of Industry
147 North Ford St.
Stellarton, NS B0K 1S0
Phone: 902-755-5425
industry@gov.ns.ca

Washington press (2), several platen job presses, lever/press

Prince Edward Island

Prince Edward Island Museum and Heritage Foundation
2 Kent Street
Charlottetown, PEI C1A 1M6
Phone: 902-368-6600
peimuse@pei.sympatico.ca
www.peimuseum.com

Quebec

Brome County Historical Society Museum
130 Lakeside
Knowlton, QC J0E 1V0
Phone: 450-243-6782
bchs@endirect.qc.ca

platen job press

Canadian Museum of Civilization
100 Laurier Street,
Gatineau, QC J8X 4H2
Phone: 819-776-7000
www.civilization.ca

platen job press and other presses

Colby-Curtis Museum/ Stanstead Historical Society
535 Dufferin
Stanstead, QC J0B 3E0
Phone: 819-876-7322
mccrcip@interlinx.qc.ca

Eberdt Museum of Communications / Musée historique des communications
30 Principale Sud Sutton,
Sutton, QC J0E 2K0
Phone: 514-538-2649

Ontario

*Black Creek Pioneer Village
1000 Murray Ross Parkway
Toronto, ON M3J 2P3
Phone: 416-736-1733
www.trca.on.ca/parks_and_culture/places_to_visit/black_creek

Washington press, Tufts press, platen job press, hand cylinder press

Brant Museum and Archives
57 Charlotte Street
Brantford, ON N3T 2W6
Phone: 519-752-2483
bcma@bfree.on.ca
www.brantmuseum.ca

Gordon Franklin press and other presses

Cumberland Heritage Village Museum
2940 Old Montreal Rd.,
Cumberland, ON K0A 1S0
Phone: 613-833-3059
museums@ottawa.ca

cylinder press

Doon Heritage Crossroads
10 Huron Road at Homer Watson Boulevard
Kitchener, ON N2P 2R7
Phone: 519-748-1914
www.region.waterloo.on.ca

Washington press, platen job press

*Fanshawe Pioneer Village (London & Middlesex Heritage Museum)
2609 Fanshawe Park Rd. E
London, ON N5X 4A1
Phone: 519-457-1296
info@fanshawepioneervillage.ca

Washington press

Fenelon Falls Museum
50 Oak Street
Fenelon Falls, ON K0M 1N0
Phone: 705-887-1044
maryboro2003@yahoo.ca

platen job press

Haileybury Heritage Museum
575 Main St.
Haileybury, ON P0J 1K0
Phone: 705-672-1922
hhmuseum@onlink.net

platen job press

Haliburton Highlands Museum
66 Museum Road
Haliburton, ON K0M 1S0
Phone: 705-457-2760
haliburtonmuseum@halhinet.on.ca

platen job press

Huron County Museum
110 North Street
Goderich, ON N7A 2T8
Phone: 519-524-2686
www.huroncounty.ca/museum

*Lang Pioneer Village
104 Lang Road
Keene, ON
c/o 470 Water Street
Peterborough, ON K9H 3M3
Phone: 705-295-6694 or 1-866 289-5264
www.langpioneervillage.ca

Washington press, platen job press

*Mackenzie House Museum
82 Bond Street
Toronto, ON M5B 1X2
Phone: 416-392-6915
www.Toronto.ca/culture/Mackenzie_house.htm

Washington press

*Mackenzie Printery and Newspaper Museum
1 Queenston Street
Queenston, ON L0S 1L0
Phone: 905-262-5676
www.mackenzieprintery.ca

Louis Roy common press, Albion press, platen job presses, linotype, monotype

Museum of London
421 Ridout Street North
London, ON N6A 5H4
Phone: 519-661-0333
mbaker@museumlondon.ca

Washington press, platen job presses (2), Columbian press

Providence Manor Sisters of Providence of St. Vincent de Paul
1200 Princess Street
Kingston, ON K7M 4S1
Phone: 613-544-4525
communications@providence.ca

platen job press, linotype

Scugog Shores Historical Museum, Village and Archives
16210 Island Road
Port Perry, ON L9L 1B4
Phone: 905-985-3589

platen job press

Stratford-Perth Museum
270 Water Street
Stratford, ON N5A 3C9
Phone: 519-271-5311
spmuseum@cyg.net

Washington Press, Excelsior press

*Strathroy Museum
34 Frank St.
Strathroy, ON N7G 3A5
Phone: 519-245-0492
info@strathroymuseum.ca

Washington press

*Upper Canada Village
13740 County Road 2
Morrisburg, ON K0C 1X0
Phone: 613-543-4328 or 1-800-437-2233
www.uppercanadavillage.com

Washington press (2), platen job press (2)

*Uxbridge Historic Centre
Box 1301
7239 Concession 6
Uxbridge, ON L9P 1N5
Phone: 905-852-5854
museum@town.uxbridge.on.ca

platen job press

*Westfield Heritage Village
1049 Kirkwall Rd.
Rockton, ON L0R 1X0
Phone: 519-621-8851 or 1-800-883-0104
westfield@speedway.ca

Washington press, platen job press, Hoe cylinder press

MANITOBA

*Crystal City Community Printing Museum
218 Broadway St S
Crystal City, MB R0K 0N0
Phone: 204-873-2293

Manitoba Agricultural Museum
3 km south of the junction of highways 1 and 34.
Box 10
Austin, MB R0H 0C0
Phone: 204-637-2354
www.ag-museum.mb.ca

*Mennonite Heritage Village Museum
231 PTH 12N
Steinbach, MB R5G 1T8
Phone: 204-326-9661

platen job press

SASKATCHEWAN

1910 Boomtown
2610 Lorne Avenue South
Saskatoon, SK S7J 0S6
Phone: 306-931-1910
saskatoon@wdm.ca

platen job press, hand cylinder press, Washington [Reliance] press

Evolution of Education Museum
2nd Avenue West and Marquis Road,
Prince Albert, Saskatchewan S6V 8A9

Phone: 306-764-2992
historypa@citypa.com

platen job press

Heritage Farm and Village
Box 183, Highways 16 & 40
North Battleford, SK S9A 2Y1
Phone: 306-445-8033
nbattleford@wdm.ca

Washington press

*Sukanen Ship Pioneer Village & Museum
Hwy 2 South
Moose Jaw, SK S6H 0A5
Phone: 306-693-7315
pmjohnson@Sasktel.net

platen job press and other presses

ALBERTA

*Alberta Heritage Park Historical Village
12845 - 102nd Avenue NW
Calgary, AB T2V 2X3
Phone: 403-268-8500
info@heritagepark.ab.ca

cylinder press, platen job press and other presses

Royal Alberta Museum
12845 - 102nd Avenue
Edmonton, AB T5N 0M6
Phone: 780-453-9100
www.royalalbertamuseum.ca

BRITISH COLUMBIA

*Barkerville Historic Town
14301 Barkerville
Wells, BC V0K 1B0
Phone: 250-994-3332
www.barkerville.ca

platen job presses, Washington press

Fort Steele Heritage Town
9851 Highway 93/95
Fort Steele, BC V0B 1N0
Phone: 250-417.6000
info@FortSteele.bc.ca
www.fortsteele.ca

*Greater Vernon Museum and Archives

3009 - 32nd Avenue
Vernon, BC V1T 2L8
www.vernonmuseum.ca
Phone: 250-542-3142

platen job press, modern cylinder press, linotype

Royal British Columbia Museum

675 Belleville Street
Victoria, BC V8W 9W2
Phone: 250-356-7226
www.royalbcmuseum.bc.ca

Bishop Demers press, Columbian press

Yukon

MacBride Museum

1st Avenue & Wood Street
Whitehorse, YT Y1A 1A4
Phone: 867-667-2709
info@macbridemuseum.com

platen job press

Historic Sites and Museums with letterpress equipment in their collections, but not on permanent display

Canadian Science and Technology Museum

1867 St. Laurent Blvd.
Ottawa, ON K1G 5A3
Phone: 613-991-3044
www.sciencetech.technomuses.ca

wide range of presses

Elgin County Pioneer Museum

32 Talbot Street
St. Thomas, ON N5P 1A3
Phone: 519-631-6537
ecpm@elgin-county.on.ca

limited collection of printing equipment

Humboldt and District Museum and Gallery

Main Street and Sixth Avenue
Humboldt, Saskatchewan S0K 2A0
Phone: 306-682-5226
humboldt.museum@sasktel.net

platen job press, linotype

Welland Historical Museum

140 King Street
Welland, ON L3B 3J3
Phone: 905-732-2215
wellandhistoricalmuseum@cogeco.net

limited collection of printing equipment

Other Public Institutions with extensive collections of letterpress equipment

Dawson Collection

Nova Scotia College of Art and Design
5163 Duke Street
Halifax, NS B3J 3J6
Phone: 902-444-9600

wide range of presses; replica wooden common press

Robertson Davies Library

Massey College, University of Toronto
4 Devonshire Place
Toronto, ON M5S 2E1
Phone: 416-978-1759

wide range of presses

Notes

Because so little printing equipment used in Canada in the earliest years survives, a number of the photographs in this book are anachronistic, using presses and types of a later time. Nevertheless, the descriptions of the basic printing processes, if not the actual tools, are essentially accurate.

Below are the sources of the quotations in the text and, in two instances, for more information. For bibliographic details not listed in full, see the Selected Bibliography.

Introduction

"By means of the press" — Johnson, *Typographia or the Printers' Instructor*, p. 78.

1: King's Printers

William Brown's finances — His records from 1764 to 1800 are available in an appendix to Fleming and Alston, *Early Canadian Printing;* "A printer is indispensably necessary," — Simcoe in E.A. Cruickshank, (ed.), *The Correspondence of Lieut. Governor John Graves Simcoe.* Toronto: Ontario Historical Society, 1921-31. Vol. 1, p. 48; "the most effectual Means" and "have nothing so much at heart" — *Quebec Gazette/La Gazette de Québec*, 1764 June 21; "to mention my name without apprising me" — Osgoode in William Colgate, "Louis Roy: First Printer in Upper Canada" *Ontario History Papers and Records*, Vol. XLII, No. 3 (1951); "cannot write good English" — Elizabeth Simcoe in J. Ross Robertson, *The Diary of Mrs. John Graves Simcoe.* Toronto: William Briggs, 1911 (reprinted Toronto: Coles Publishing Co., 1973), p.161; "We are put to great Inconvenience" — Osgoode in Colgate, "Louis Roy;" "His political principles" — The *Argus* in Colgate, "Roy, Louis."

2: Independent Printer-Editors 1800-1850.

"own good sense and discretion" — Littlehales in Cruickshank, The Correspondence ... Simcoe, vol. 3, p. 346; "unprincipled and unattached republicans" — Elmsley in E.A. Cruikshank, (ed.), *The Correspondence of the Honourable Peter Russell*, Toronto: *Ontario Historical Society*, 1935, Vol. 11, p. 103-04; "treasonable or seditious conduct" and "As the people's printer" — Silvester Tiffany to Peter Russell, 1798 April 30 and May 13. *Russell Papers*, Toronto Reference Library; "[Despite] his enjoyment" — *Colonial Advocate* 1824 May 18; "To become an advocate" — *Colonial Advocate* 1824 July 01; "Up then, and be doing" — *Colonial Advocate* 1824 July 08; "[T]he true friend to his country" — *Colonial Advocate* 1826 May 03; "The Constitution, the whole Constitution," — *Novascotian* 1828 January 02; "We have here twelve men" — *Novascotian* 1830 May 20; "I conjure you" — *Novascotian* 1835 March 15; "I am awkward" — *Colonial Advocate* 1832 June 26; "My dear Susan Ann" — Howe in M.G. Parks, editor. *My Dear Susan Anne: Letters of Joseph Howe to His Wife 1829-1836*. St. Johns: Jesperson Press, 1985, pp. 18-19; "The proprietor and editor" — *Colonial Advocate* 1826 April 27.

3: Community Printers

"It is usual" — Brown in J.M.S. Careless, *Brown of the Globe, volume One: The Voice of Upper Canada 1818-1859*, Toronto: Macmillan, 1959, p. 42; "duplicity is sufficient" — Bibb in Jane Rhodes, *Mary Ann Shadd Carey: The Black Press and*

Protest in the Nineteenth Century, Bloomington, IN: Indiana University Press, 1998, p. 73; "None of the papers published" — Shadd in Rhodes, p. 85; "Independence in Everything" — Motto of the Creemore *Advertiser*, 1886 March 11; "a crucial difference" — Eli Maclaren "Resistance, Rebellion and Print in the Northwest, in Yvan Lamonde et al, *History of the Book in Canada Volume Two 1840-1918*. Toronto: University of Toronto Press, 2005, p. 343; "We intend" "We shall ever" and "He alone is responsible" — De Cosmos in George Woodcock, *Amor De Cosmos Journalist and Reformer*. Toronto: Oxford University Press, 1975, pp. 34-36; "Governments, corporations" — Unidentified writer in Woodcock, p. 168; The Bishop Demers press — Jeremiah Saunders, "The Fraser River Gold Rush of 1858 and the Victoria Newspaper Boom," unpublished paper for the University of British Columbia School of Library, Archival and Information Studies, 2005.

Appendix A: Historically Significant Hand Presses

"to render the hand press efficient" — Ramage in "About the Ramage Screw Press," Briar Press, www.briarpress.org.

Appendix B: Ink and Inking

"Printer's Red Ink - Soft varnish and vermilion" — *Colonial Advocate* 1826 April 06.

Selected Bibliography

For more information about the lives of the printers referred to in the text, see their entries in the *Dictionary of Canadian Biography*, Toronto: University of Toronto Press.

Bassam, Bertha, *The First Printers and Newspapers in Canada*. Toronto: University of Toronto School of Library Science Monograph Series in Librarianship, No. 1, 1968.

Benn, Carl, "The Upper Canadian Press, 1793-1815." *Ontario History*, Vol. LXX, No. 2, June 1978.

Blyth, J. A., "The Development of the Paper Industry in Old Ontario, 1824-1867." *Ontario History*, Vol. LXII, No. 2, June 1970.

Comparato, Frank E, *Chronicles of Genius and Folly: R. Hoe & Company and the Printing Press as a Service to Democracy*. Culuver City, CA, 1978.

Dewalt, Bryan, *Technology and Canadian Printing: A History from Lead Type to Lasers*. Ottawa: National Museum of Science and Technology, 1995.

Fauteau, Aegidius, *The Introduction of Printing into Canada*. Montreal: Rolland Paper Company, 1930.

Fetherling, Douglas, *The Rise of the Canadian Newspaper*. Toronto: Oxford University Press, 1990.

Fiftieth Anniversary Committee, *Canadian Book of Printing: How Printing Came to Canada and the Story of the Graphic Arts, Told Mainly in Pictures*. Toronto: Toronto Public Libraries, 1940.

Fleming, Patricia Lockhart & Sandra Alston, *Early Canadian Printing: A Supplement to Marie Tremaine's A Bibliography of Canadian Imprints 1751-1800*. Toronto: University of Toronto Press, 1999.

Fleming, Patricia Lockhart, *Atlantic Canadian Imprints, 1801-1820: A Bibliography*. Toronto: University of Toronto Press, 1991.

Fleming, Patricia Lockhart, Gilles Gallichan, and Yvan Lamonde, editors, *History of the Book in Canada: Volume One, Beginnings to 1840*. Toronto: University of Toronto Press, 2004.

Fleming, Patricia Lockhart, *Upper Canadian Imprints, 1801-1841: A Bibliography*. Toronto: University of Toronto Press, 1988.

Gundy, H, Pearson, *The Spread of Printing, Western Hemisphere, Canada*. Amsterdam: Vangendt & Co., 1972.

Gundy, H. Pearson, *Early Printers and Printing in the Canadas*. Toronto: Bibliographical Society of Canada, 1957.

Hamilton, Milton W., *The Country Printer: New York State, 1785-1830, 2nd edition* [1936]. Port Washington, NY: Ira J. Friedman, Inc. 1964.

Haworth, Eric, *Imprint of a Nation*. Toronto: Baxter Publishing, 1969.

Hudson, Graham, *The Victorian Printer: Shire Album 329*. Princes Risborough, Buckinghamshire: Shire Publications, 1996.

Johnson, J., *Typographia, or the Printers' Instructor: [Vol. 1 & 2]* ...[1824] London: Gregg Press, 1966.

Kesterton, W. H., *A History of Journalism in Canada*. Ottawa: Carleton University Press, 1964.

Lamonde, Yvan, Patricia Lockhart Fleming, and Fiona A, Black, editors, *History of the Book in Canada: Volume Two, 1840-1918*. Toronto: University of Toronto Press, 2005.

Martel, J.S. "The Press of the Maritime Provinces in the 1830s." *Canadian Historical Review*, Vol. XIX, 1938.

Moran, James, *Printing Presses: History & Development from the 15th Century to Modern Times*. Berkeley: University of California Press, 1978.

Parker, George L., *Beginnings of the Book Trade in Canada*. Toronto: University of Toronto Press, 1985.

Raible, Chris, *A Colonial Advocate: The Launching of his Newspaper and the Queenston Career of William Lyon Mackenzie*. Creemore, ON: Curiosity House, 1999.

Raible, Chris, *Muddy York Mud: Scandal & Scurrility in Upper Canada*. Creemore, ON: Curiosity House, 1992.

Raible. Chris, "'A Printer is Indispensably Necessary': The tribulations of Canada's earliest printers." *The Beaver*, August-September 1997.

Rutherford, Paul, *The Making of the Canadian Media.* Toronto: McGraw-Hill Ryerson, 1978.
Saxe, Stephen O., *American Iron Hand Presses: The Story of the Iron Press in America*. New Castle, DE: Oak Knoll Press, 1992.
Stabile, Julie, "The Economics of an Early Nineteenth-Century Toronto Newspaper Shop." *Papers of the Bibliographical Society of Canada*, Vol. 41, #1, Spring 2003.
Steinberg, S.H., John Trevitt, *Five Hundred Years of Printing, New Edition*. Newcastle, DE: Oak Knoll Press & British Library, 1996.
Thomas, Isaiah, *The History of Printing in America* [1810]. New York: Weathervane Press, 1970.
Tobin, Brian, *The Upper Canada Gazette and Its Printers.* Toronto: Ontario Legislative Library, 1993.
Tremaine, Marie, *A Bibliography of Canadian Imprints 1751-1800* [1952]. Toronto: University of Toronto Press, 1999.
Wallace, W. Stewart, "The First Journalists in Upper Canada." *Canadian Historical Review,* Vol. 26, No. 4, December 1945.
Wallace, W. W. "The Periodical Literature of Upper Canada." *Canadian Historical Review,* Vol. 13, No. 1, March 1931.
Whitelaw, Marjory, *First Impressions: Early Printing in Nova Scotia.* Halifax: Nova Scotia Museum, 1987.
Wroth, Lawrence C., *The Colonial Printer* [1931]. New York: Dover Publications, 1994.
Zerker, Sally F., *The Rise and Fall of the Toronto Typographical Union 1832-1972: A Case Study of Foreign Domination.* Toronto: University of Toronto Press, 1982.

Photo Credits

The publisher wishes to thank the curators and interpretive staff at the participating sites for supplying photographs from their own collections and for their co-operation and help during the photo shoots.

The images on the following pages were photographed by Rob Skeoch at Black Creek Pioneer Village, Toronto and Region Conservation Authority: 3, 5, 12, 17, 18, 19T, 31, 41, 51, 57, 58, 59, 67, 75, 76

The images on the following pages were photographed by Rob Skeoch at Mackenzie House: Back Cover Centre, 2, 15, 22, 24L, 26, 37R, 40L, 42, 43, 46, 47, 49, 63, 66, 73

The image on the following page was photographed by Rob Skeoch at The Mackenzie Printery and Newspaper Museum: 9

The images on the following pages were photographed by Jackie MacRae of Behold Photographics at Upper Canada Village: Front Cover, Back Cover Left, Back Cover Top, Back Cover Bottom, 10, 11, 13, 14, 16, 19R, 20, 21, 23, 24T, 25, 27, 28, 32, 33, 35, 37L, 38, 40T, 44, 45, 48, 52, 53, 54, 55, 56TR, 61, 64, 65, 68, 69, 72, 74, 77, 78

Other photographs were supplied by and appear courtesy of:

Lang Pioneer Village Museum: 8, 39, 62, 71

Village Historique Acadien: 36, 70

Photo Fredericton: 56TL

R=right; L=left; T=top

Index